BIOLOGY PRACTICAL GUIDE 6

DEVELOPMENT, CONTROL, AND INTEGRATION

Revised Nuffield Advanced Science
Published for the Nuffield–Chelsea Curriculum Trust
by Longman Group Limited

Longman Group UK Limited
Longman House, Burnt Mill, Harlow, Essex, CM20 2JE, England
and Associated Companies throughout the world.

First published 1970
Revised edition first published 1986
Fifth impression 1992
Copyright © 1970, 1986, The Nuffield-Chelsea Curriculum Trust

Design and art direction by Ivan Dodd
Illustrations by Oxford Illustrators

Set in Times Roman and Univers
Produced by Longman Singapore Publishers (Pte) Ltd
Printed in Singapore

ISBN 0 582 35510 9

Cover photograph
A section through the thyroid gland ($\times 600$). See investigation 24A,
'The microscopic structure of endocrine glands'.
Photograph, Biophoto Associates.

CONTENTS

SAFETY

In these *Practical guides*, we have used the internationally accepted signs given below to show when you should pay special attention to safety.

 highly flammable

 take care! (general warning)

 explosive

 risk of electric shock

 toxic

 naked flames prohibited

 corrosive

 wear eye protection

 radioactive

 wear hand protection

INTRODUCTION

The practical investigations in this *Guide* relate to the topics covered in *Study guide II*, Part One, 'Inheritance and development', Chapters 22 to 25. Cross references to the *Study guide* are given.

METHODS OF REPRODUCTION

Investigation 22A Reproduction in fungi. (*Study guide* 22.2 'Asexual reproduction' and 22.6 'Gametogenesis and fertilization'.)
Three species of fungi are investigated microscopically for evidence of the basic processes involved in sexual and asexual reproduction.

Investigation 22B Reproduction in *Marsilea vestita*, the shamrock fern. (*Study guide* 22.5 'Variations in life cycles' and 22.6 'Gametogenesis and fertilization'.)
Germination of *Marsilea* sporocarps is initiated and subsequent events are followed microscopically.

Investigation 22C The life cycle of *Dryopteris filix-mas*, the male fern. (*Study guide* 22.5 'Variations in life cycles'.)
The nature of reproduction in *Dryopteris* is investigated.

Investigation 22D Pollination mechanisms in angiosperms. (*Study guide* 22.6 'Gametogenesis and fertilization'.)
The mechanisms that plants possess to promote pollination are examined, and features characteristic of a particular pollination mechanism identified.

Investigation 22E Events leading to fertilization in angiosperms. (*Study guide* 22.5 'Variations in life cycles' and 22.6 'Gametogenesis and fertilization'.)
The development of the pollen tube is investigated in *Pelargonium* and *Impatiens*. Gametogenesis in angiosperms and mammals is compared.

Investigation 22F The structure and function of the mammalian gonads. (*Study guide* 22.6 'Gametogenesis and fertilization'.)
Microslides of sections through the mammalian gonads are studied.

Chapter 23   THE NATURE OF DEVELOPMENT

Investigation 23A Tracing the early development of a nematode worm, *Rhabditis*. (*Study guide* 23.2 'Early development'.)
Changes in the eggs of *Rhabditis* during development are observed.

Investigation 23B Embryonic development in shepherd's purse, *Capsella bursa-pastoris*. (*Study guide* 23.2 'Early development'.)
Developing embryos of shepherd's purse are examined and compared with that of *Rhabditis*.

Investigation 23C Morphogenesis in amphibians. (*Study guide* 23.3 'Development of the body plan'.)
Three-dimensional models are constructed, based on a series of two-dimensional fate maps, to show the movement of cell masses during the early development of a *Xenopus* embryo.

Investigation 23D Growth and development of the fore-limb of mice. (*Study guide* 23.5 'Growth' and 23.8 'Pattern formation in the development of limbs'.)
The growth and development of the fore-limb of mice are investigated by measuring limb elements in series of photographs of fore-limbs of different ages.

Investigation 23E Growing new plants from old. (*Study guide* 23.9 'The pattern of plant and animal development compared'.)
A simple method of tissue culture is used to grow a new plant from a small part of the mature one.

Chapter 24   CONTROL AND INTEGRATION THROUGH THE INTERNAL ENVIRONMENT

Investigation 24A The microscopic structure of endocrine glands. (*Study guide* 24.1 'Endocrine communication and control in animals' and 24.2 'Hormone synthesis and release'.)
A microscopic examination reveals some of the features of hormone secreting glands.

Investigation 24B The effect of IAA on the growth of coleoptiles and radicles. (*Study guide* 24.7 'Communication and control in plant growth'.)
The response of plant coleoptiles and radicles to indole-3-ethanoic(acetic) acid (IAA) is investigated.

Investigation 24C Stimulation of amylase production in germinating barley grains. (*Study guide* 24.7 'Communication and control in plant growth'.)
This investigation demonstrates how the food reserves are made soluble for transport to the embryo.

Investigation 24D The effect of plant hormones on seed germination. (*Study guide* 24.7 'Communication and control in plant growth'.)
This investigation demonstrates the role of plant hormones in dormancy.

Chapter 25   **DEVELOPMENT AND THE EXTERNAL ENVIRONMENT**

Investigation 25A The effect of temperature on root growth. (*Study guide* 25.2 'The external environment in relation to growth and development'.)
Pea seedlings are used to study the effect of temperature change on root growth.

Investigation 25B Effects of light on the germination of lettuce seeds. (*Study guide* 25.3 'Light and plant growth'.)
The nature of light is outlined and its role in germination of seeds investigated in a series of experiments with lettuce seeds.

Investigation 25C Early environment and later behaviour of mice. (*Study guide* 25.4 'The study of ontogeny'.)
The effect of one type of change in early environment on later behaviour is investigated using mice.

A note for users of this *Practical guide*

The instructions given for the investigations are intended for use as guidelines only. We hope that you will modify and extend the techniques that have been described to meet your own requirements. Other organisms should certainly be tried, depending on what is most readily available. Some of these investigations may lend themselves to further work in a Project.

It may not always be possible, for various reasons, for you to do a practical investigation for yourself. A study of data from another source is perfectly acceptable in such a case.

METHODS OF REPRODUCTION

All individuals have a finite life-span, and for any species to survive, replacement by reproduction must take place. Despite the variety of methods of reproduction, all have one feature in common, the inheritance from the previous generation of a genetic complement for the next.

Asexual reproduction depends entirely upon cell division by mitosis; sexual reproduction also involves two unique events, cell division by meiosis, and fertilization. In both sexual and asexual reproduction there are mechanisms to keep constant the amount of genetic material in the new individual. If we think of transfer of genetic material as the essence of reproduction, then the structures and processes linked with the handing on of the genetic material can be seen as mechanisms to aid its success.

INVESTIGATION
22A   Reproduction in fungi

(*Study guide* 22.2 'Asexual reproduction' and 22.6 'Gametogenesis and fertilization'.)

The fungi are successful organisms, both in terms of their distribution and of their abundance. They produce spores that can be found in almost every habitat. This success can be related to their mechanisms of reproduction and dispersal. You are going to investigate three species of fungi for evidence of the basic processes involved in asexual and sexual reproduction. You will study a unicellular fungus *Saccharomyces cerevisiae*, baker's yeast, and two filamentous fungi, *Penicillium expansum*, a blue-mould, and *Mucor hiemalis*, a form of pin-mould (*figure 1*). The last fungus is found in two types, or strains, called the + strain and the − strain.

Take care to follow the recommended safety precautions when handling micro-organisms, and be careful to avoid dispersing spores around the laboratory.

Figure 1   (*Opposite*)
Three types of fungi.
a *Saccharomyces* (yeast) cells (× 1150).
b *Penicillium* conidiophores (× 500).
c *Mucor* mycelium (× 50).
Photographs: **a** and **c** *Biophoto Associates;* **b** *Commonwealth Mycological Institute.*

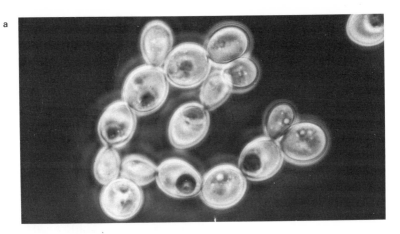

a

b

c

Saccharomyces cerevisiae

Yeast must be one of the first micro-organisms to have been used by humans – an early example of biotechnology. Under optimum conditions yeast cells reproduce very rapidly in culture, doubling their numbers in less than thirty minutes. We might, therefore, expect to find evidence of reproduction in a culture that has been kept under suitable conditions.

Procedure
1   Place a drop of the culture of S. *cerevisiae* on a microscope slide and add a coverslip.
2   Observe the yeast under high power (× 400) for any signs of reproduction. You may find it helpful to keep just one yeast cell under observation from time to time over a period of 15 to 20 minutes, but switch off the light between observations to avoid drying out the preparation.
3   Make a record of what you see. Annotated drawings are a suitable way of doing this.

Penicillium expansum

This is a species of the genus from which the first antibiotic, penicillin, was extracted. *Penicillium* moulds are very commonly found on rotting food, such as fruit, so they must have efficient methods of both reproduction and dispersal.

Procedure
1   Collect an agar plate on which a culture of *Penicillium expansum* is growing.
2   Observe the mould through a low power stereo binocular microscope (× 10 and × 20) for evidence of reproduction.
3   Gently remove a small portion of the filamentous mass with forceps. Mount it on a microscope slide and add a coverslip.
4   Under the low and high power of your microscope, make more detailed observations of the structures you find.

Mucor hiemalis

When bread goes mouldy, *Mucor* is one of the commonest moulds to colonize such a habitat. It forms black patches on the bread, which, when observed with a hand lens, look just like a mass of dark pin-heads – hence its common name, pin-mould.

Procedure

1 Collect an agar plate which has been inoculated with two strains of
  Mucor hiemalis (+ and −).
2 Observe the mould more closely through a low power stereo binocular
  microscope. Look for possible reproductive structures. Look closely at
  the strip of fungal mycelia where the two strains have met.
3 Mount representative parts of the strip on a microscope slide, and add
  a coverslip.
4 Examine the slide under high power, look for any signs of interaction
  between the two strains, and make annotated sketches to record your
  observations.

Questions

a *The simplest form of asexual reproduction is called budding. This
  process is found in* **Saccharomyces cerevisiae.** *What nuclear and
  cytoplasmic processes must occur before a new organism is formed
  by this process?*

b *Asexual reproduction in* **Penicillium expansum** *follows the same basic
  process, but with structural elaborations. What modifications did
  you observe?*

c *What are the advantages of the two types of asexual reproduction?*

d *What differences were you able to detect between the two* **Mucor**
  strains?

e *What evidence did you observe of sexual reproduction in* **Mucor***?
  Form a hypothesis concerning the occurrence of the sexual process in
  this mould. Design an experiment that would test it.*

f *Why do you think that the two types of* **Mucor** *are named + and −
  strains, and not male and female?*

g *Sexual and asexual structures are quite distinct in* **Mucor***. How may
  their functions differ? How might this be related to the dispersal of
  the mould and to factors that adversely affect its survival?*

h *Draw a diagram showing the life cycle of* **Mucor hiemalis,** *including
  both sexual and asexual phases (you may need to consult suitable
  texts). Why do you think that scientists are still not certain of
  exactly when meiosis takes place in the life cycle?*

i *Why must meiosis occur during the life cycle of a sexually
  reproducing organism?*

INVESTIGATION

22B Reproduction in *Marsilea vestita*, the shamrock fern

(*Study guide* 22.5 'Variations in life cycles' and 22.6 'Gametogenesis and fertilization'.)

Marsilea is a pteridophyte and species are found in North America and Europe, but it is not native to the United Kingdom. It is also called the shamrock fern because of the shape of the leaf (*figure 2*). *Marsilea* grows on mud in shallow ponds, marshes, and ditches. The plant spreads by means of a rhizome, which may be completely submerged, and the plant may cover up to 25 square metres. Under suitable conditions, possibly when the environment temporarily dries up, *Marsilea* produces reproductive structures called sporocarps. These sporocarps possess a very resistant wall and under normal conditions they probably do not germinate for some years. It is known that under laboratory conditions

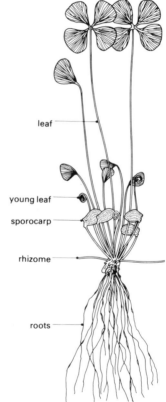

leaf

young leaf

sporocarp

rhizome

roots

Figure 2
Marsilea vestita, the shamrock fern.
After Benson, L., Plant classification, *D. C. Heath, 1959.*

they remain viable for over 30 years. On germination, spores are released, from one type of which motile male gametes are produced. In this investigation you are going to initiate the germination of the sporocarp and follow the course of events.

Procedure
1   Take one or two sporocarps and, using a scalpel, carefully remove about 1 mm from one end of the sporocarp.
2   Place the sporocarp in a Petri dish half-filled with distilled water. Note the time.
3   Observe the sporocarp through a low power stereo binocular microscope (× 20), every ten minutes for the next hour.
4   Record the events during spore release.
5   Use a pipette to transfer a sample of the materials that you see to a microscope slide for further examination under high power.
6   Record your observations.
7   Place the cover on the Petri dish and leave it for a further period of about 14 hours.
8   Remove some of the spores and examine them on a microscope slide under high power. Record any differences compared with your previous observations.
9   At 16 hours, or a little later, active male gametes, called antherozoids are released. Observe the spores for antherozoid activity. If you spot any, try to make out their general structure. Make a drawing and compare it with the different types of sperm shown in *figure 3*.
10 Over the next few weeks keep the dish covered and topped up with water. Examine it from time to time for any signs of embryonic developments. Make a record of any that take place.

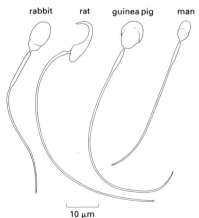

Figure 3
Different types of sperm.
After Young, J. Z., The life of mammals, *2nd edition, Oxford University Press, 1975.*

Questions

a *How many types of spore does* Marsilea vestita *produce? How do they differ?*

b *What is the mechanism of spore release? Suggest the environmental conditions that are necessary for successful release and fertilization.*

c *What events take place immediately before fertilization? What mechanisms may help to increase the chances of gametes meeting?*

d *What are the advantages and disadvantages of organisms releasing their gametes directly into their surroundings?*

e *Based upon your observations of early development, how does the embryo obtain the nutrients needed for growth and development?*

INVESTIGATION

22C The life cycle of *Dryopteris filix-mas*, the male fern

(*Study guide* 22.5 'Variations in life cycles'.)

Dryopteris is typical of what we normally think of as a fern. It is, apart from *Pteridium aquilinum*, bracken, possibly the commonest type of fern found in the United Kingdom. The name, male fern, is believed to have been given the species because of the robust upright growth of the frond. All ferns exist in two quite distinct forms during their life cycle, each with a radically different morphology. This investigation is to study the nature of reproduction in this species and to consider how this type of life cycle may affect its distribution.

Procedure

1 Take a mature fern gametophyte, or prothallus, and mount it in water in a Petri dish, so that the underside, which produces hair-like rhizoids, is uppermost.

2 Study this, using a low power stereo binocular microscope, and look for any signs of reproductive organs. Draw the prothallus to show its general structure and indicate where the reproductive organs occur.

3 Remove from the prothallus portions containing the male organs, antheridia, and the female organs, archegonia. Mount these in water on a microscope slide, add a coverslip, and examine under low and high power. If you are lucky and have mature antheridia, you may see some motile antherozoids. Supplement your observations by use of prepared slides of sections through a prothallus showing the reproductive organs.

4 Take a mature frond from the sporophyte. It is divided into a series of pinnae on either side of the axis. Each pinna is subdivided into a

a

indusium

sporangia

b

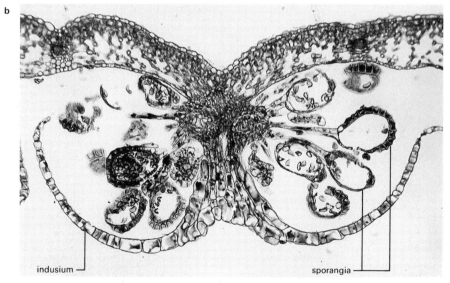

indusium

sporangia

Figure 4
a A scanning electronmicrograph of a fern sorus showing the indusium and sporangia
(× 45).
b A vertical section through a mature sorus of *Dryopteris* (× 150).
Photographs, Biophoto Associates.

number of pinnules. Examine the underside of one of these for spore bearing structures (*figure 4a*). Using a mounted needle, remove the umbrella-like indusium to expose the spore capsules, or sporangia (*figure 4b*).

5   Mount some mature sporangia, from a frond where the indusia have shrivelled up, in water on a microscope slide. Add a coverslip, and examine under low and high power.

6   Mount some more mature sporangia on a dry microscope slide and leave for a few minutes under a bright light. Keep these under observation and record any differences compared with the sporangia mounted in water.

7   Either sow some of the spores from a mature frond onto a suitable substrate, or follow the development of the prothalli you used earlier. Look for the development of a sporophyte from the prothallial tissue.

Questions

a   *In what ways is the fern gametophyte adapted to lead an independent existence?*

b   *How are the antheridia and archegonia distributed on the gametophyte?*

c   *How might the structure of the gametophyte affect the distribution of the fern?*

d   *Given that the gametophyte is haploid, by what cytological process are gametes produced?*

e   *By what cytological process does the sporophyte produce spores?*

f   *What evidence must have been obtained by scientists in order for them to understand the nature of this type of life cycle? Why is it called alternation of generations?*

Sexual reproduction in angiosperms

Angiosperms, the flowering plants, are considered to be more successful in most habitats than the pteridophytes, because there are a far greater number of species and they are able to exploit a much wider range of habitats. A major factor in this is that, in their life cycle, they are more independent of free water for fertilization than are ferns.

A feature of sexual reproduction in the angiosperms is the production and transfer of pollen. The following investigations study the sexual reproductive system of angiosperms and consider to what extent the success of the flowering plants is due to these mechanisms.

INVESTIGATION
22D Pollination mechanisms in angiosperms

(*Study guide* 22.6 'Gametogenesis and fertilization'.)

Most flowering plants produce hermaphrodite flowers and a variety of mechanisms are found that prevent or reduce the likelihood of self-pollination. The aim of this investigation is, firstly, to study the mechanisms that plants possess to achieve pollination and, secondly, to identify the features characteristic of a particular pollination mechanism.

Procedure

1   Collect flowering shoots of any species available.
2   Examine several flowers of each type carefully, using a hand lens and/or a low power stereo binocular microscope. Decide for each flower whether the anthers have opened to release pollen and whether the surface of the stigma has matured.
3   For each species record whether the anthers or the stigma mature first, or if they appear to mature at the same time.
4   Make sketches to show the relative positions of stigma and anthers. Decide whether the arrangement favours, or hinders cross-pollination.
5   Study the general structure of the flower for any evidence that may help you to decide whether the plant is wind-pollinated or insect-pollinated.
6   Draw up a table to compare, in the plant species that you have studied, the mechanisms that favour cross-pollination with those that favour self-pollination.
7   Prepare another table comparing the characteristics of wind-pollinated flowers with those of insect-pollinated ones.

Questions

a   *Consider the insect-pollinated plants that you have studied. What evidence can you find that a plant species is adapted to a particular type of insect pollinator?*

b   *Why may pollination between flowers not necessarily ensure cross-fertilization?*

c   *How might the reproductive system of angiosperms have contributed to their success?*

INVESTIGATION

22E Events leading to fertilization in angiosperms

(*Study guide* 22.5 'Variations in life cycles' and 22.6 'Gametogenesis and fertilization'.)

The arrival of compatible pollen on a receptive stigma is only the first of a long series of events that will result in the production of a complex reproductive propagule, the seed. In this practical we are going to investigate one of the first stages in the process – the development of the pollen tube. This normally grows down the style to the ovule, carrying with it gametic nuclei.

Procedure

1   Place a filter paper in each of four Petri dishes, moisten the paper with water, and replace the lids.

2   Take 1 cm³ of culture solution and make it up to 10 cm³ with distilled water.

3   Half the group should add 1 g of sucrose to this solution and half add 2 g, thus making 10 per cent and 20 per cent sucrose pollen medium.

4   Take four absolutely clean microscope slides and place three drops of the medium in the centre of each slide. Label two slides *Pelargonium* and two *Impatiens*.

5   Carefully remove five or six flowers from a *Pelargonium* and an *Impatiens* plant.

6   With the aid of a low power stereo binocular microscope, select two or three flowers of each type that are mature and shedding pollen.

7   Place a microscope slide under the field of view of the microscope and gently rub the point of a mounted needle over the anthers so that pollen falls onto the medium. Transfer any pollen from the needle by tapping the needle against a pair of forceps placed next to the slide.

8   Clean the needle thoroughly and repeat step 7 so that you have two slides with *Pelargonium* pollen and two with *Impatiens* pollen. Do not add a coverslip.

9   Note the time of adding pollen to the medium and place the slides in the Petri dishes. Handle the slides with great care so that the drops of pollen medium remain in the centre of the slides.

10   Use a microscope, with a magnification of × 100 and a graduated eyepiece, to observe the *Impatiens* slides over the next 15 minutes for signs of pollen germination. Between quick observations, turn the microscope lamp off and return the slide to the Petri dish.

11   Once the pollen tubes have started to grow, measure the length of four or five tubes at three minute intervals for the next 30 minutes or so.

12 After 45 minutes, observe the appearance of the pollen tube growth of the *Pelargonium*.

13 Collate the group results and work out the average growth with time for *Impatiens* in both solutions. Draw graphs to show the growth of your own pollen tubes with time; compare it with a graph of the group results, for both solutions.

Questions

a *What factors may have affected the rates of growth? Try to explain any differences in the growth rates shown in your graphs.*

b *What differences did you find in pollen tube growth between* **Impatiens** *and* **Pelargonium***? Propose a hypothesis to explain your answer, and outline a means to test it.*

c *Why is it wrong to call a pollen grain a gamete?*

d *What happens between the germination of the pollen grain and the formation of the seed? You will need to consult suitable texts.*

e *What are the advantages and the disadvantages of retaining the ovule and embryo inside the tissues of the adult?*

The angiosperm life cycle

At this point you should remind yourself of the reproductive system in mammals so that you can compare the process of gametogenesis in angiosperms and mammals. (See *Practical guide 3*, investigation 10A 'The relation of the urinary system of a mammal to other systems of the body'.)

f *Draw the life cycles of a mammal and an angiosperm, comparing the occurrence of meiotic and mitotic cell divisions.*

g *Consider the life cycles of a fern, an angiosperm, and a mammal. What comparisons can you draw between the haploid and diploid structures?*

INVESTIGATION
22F **The structure and function of the mammalian gonads**

(*Study guide* 22.6 'Gametogenesis and fertilization'.)

The mammalian gonads, the testis and ovary, not only produce gametes, but are also important sources of hormones which are concerned with

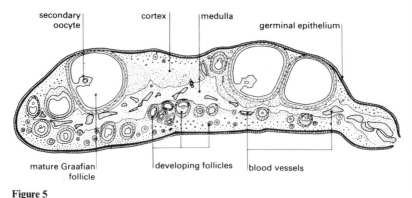

secondary oocyte | cortex | medulla | germinal epithelium

mature Graafian follicle | developing follicles | blood vessels

Figure 5
A longitudinal section of a rabbit ovary (× 7).
After Freeman, W. H. and Bracegirdle, B., An atlas of histology, *2nd edition, Heinemann Educational Books, 1967.*

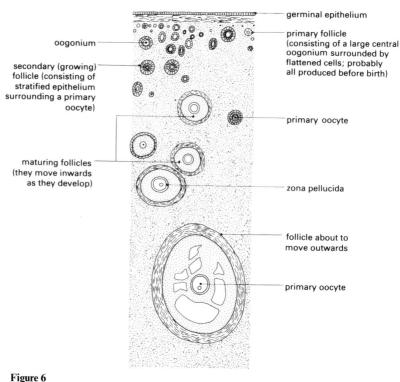

germinal epithelium

oogonium

primary follicle (consisting of a large central oogonium surrounded by flattened cells; probably all produced before birth)

secondary (growing) follicle (consisting of stratified epithelium surrounding a primary oocyte)

primary oocyte

maturing follicles (they move inwards as they develop)

zona pellucida

follicle about to move outwards

primary oocyte

Figure 6
A transverse section of part of a rabbit ovary showing follicles at different stages of development (× 175).
After Freeman, W. H. and Bracegirdle, B., An atlas of histology, *2nd edition, Heinemann Educational Books, 1967.*

16   Development, control, and integration

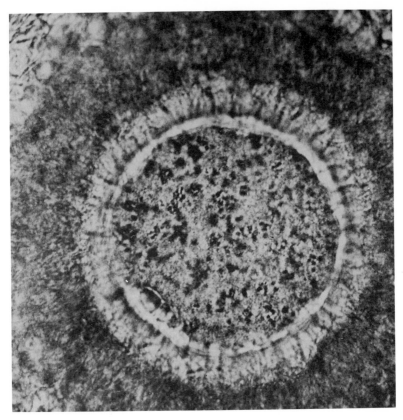

Figure 7
A human primary oocyte (× 600).
Photograph, from the film 'The first days of life' by Guigoz, distributed by Boulton-Hawker Films Ltd. Copyright Guigoz – Claude Edelmann – Jean-Marie Baufle.

reproductive development and sexual cycles. In this investigation we are going to use microslides of sections through the gonads in order to learn more about these activities.

Part 1 Ovary

Procedure

1 Set up a microscope to give you optimum illumination. Then collect a microslide of a section through a mammalian ovary.
2 Examine the preparation under low and medium power. Identify the structures shown in *figure 5*.
3 Search the preparation under medium (× 100) and high (× 400) power for different stages of follicle development as shown in *figure 6*.

4 Draw a simplified diagram of part of your preparation, as seen under high power. Annotate it to relate the process of oogenesis to the structures shown.

5 Obtain a microslide of human sperm. Measure the size of the head and tail of a sperm. Now examine *figure 7* of a mature human primary oocyte. Calculate the size of this and compare the sizes of the human egg and sperm head.

Questions

a *Examine the development of the follicles in the preparation. At what stage of the oestrous cycle was the ovary? Explain your answer.*

b *How much does the size of (1) a follicle and (2) an oocyte increase during development?*

c *How does the size of a mature human primary oocyte compare with that of the head of a human sperm?*

d *What structures in the ovary secrete hormones? Consult texts to find out. Add annotations about this to the diagram you produced in 4 above.*

Part 2 Testis

Procedure

1 After you have set up your microscope correctly, examine a microslide of a section through a testis. Draw a low power plan of the outer region of it.

2 Search at medium ($\times$ 100) and high ($\times$ 400) power and record the distribution of the most darkly staining cells.

3 Using high power, identify the structures shown in *figure 8*.

4 Measure the size of the testis and also of a mature sperm.

5 Draw a simplified diagram of a section through a seminiferous tubule, as seen under high power. Annotate your diagram to relate it to the process of spermatogenesis.

6 Assuming the testis to be a sphere, that your section is cut through the centre and is 5 µm thick, estimate the number of sperm present in the testis at any one time.

Questions

a *What is the size of the head of a mature sperm in your testis preparation? How many times longer is the tail compared with the head? Compare this result with the size of a human sperm.*

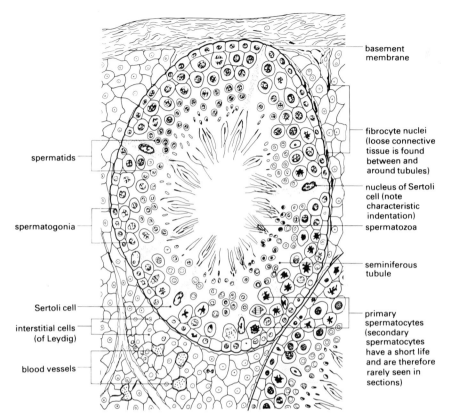

basement membrane

fibrocyte nuclei (loose connective tissue is found between and around tubules)

spermatids

nucleus of Sertoli cell (note characteristic indentation)

spermatozoa

spermatogonia

seminiferous tubule

Sertoli cell

primary spermatocytes (secondary spermatocytes have a short life and are therefore rarely seen in sections)

interstitial cells (of Leydig)

blood vessels

Figure 8
A transverse section of part of a cat testis showing seminiferous tubules (× 325).
After Freeman, W. H. and Bracegirdle, B., An atlas of histology, *2nd edition, Heinemann Educational Books, 1967.*

b *What stains were used in making the preparation? Consult texts to find out what structures they would stain.*

c *Which cells were most darkly stained? Suggest why these cells should take up more stain than others.*

d *What parts of the testis produce hormones? Consult texts to find out. Add annotations about these to the diagram you produced in 5 above.*

e *What was your estimate of the number of sperm present in the testis? Explain how you arrived at the number.*

f *Why should there be more male gametes than female?*

g *What are the similarities and differences between gametogenesis in angiosperms and mammals?*

CHAPTER 23

THE NATURE OF DEVELOPMENT

Animals and plants show differences in both the details and the sequence of events that occur during their early development. The following investigations probe the phases of early development.

INVESTIGATION
23A Tracing the early development of a nematode worm, *Rhabditis*

(*Study guide* 23.2 'Early development'.)

Round worms, phylum Nematoda, are one of the commonest forms of animal life on this planet, yet we very seldom see them. Some are free-living forms found in the soil and in aquatic environments, but many species are parasitic on a wide range of other organisms. Some species provide a suitable and readily available source of developing embryos. Earthworms are parasitized by species of nematode of the genus *Rhabditis*. The life cycle of the organism is partly parasitic and partly free-living. A larval stage, found in the earthworm, will not develop to maturity until its earthworm host dies. The larvae then feed on the decaying host and develop to mature adults. Several generations may be produced as the remains of the earthworm decay away. When the food sources are exhausted the larvae migrate and may enter and parasitize other earthworms.

Procedure

1 *Rhabditis* may be obtained from a culture or from a dead and decaying earthworm. In the latter case a piece of a dead earthworm will have been left on the surface of some soil. Remove the piece, using a pair of forceps, and gently wash it in a little water in a watch-glass.
2 Use a low power stereo binocular microscope to search for the nematodes. Larger specimens are up to 1.5 mm long.
3 Transfer some nematodes to a microscope slide with a fine pipette. Gently lower a coverslip onto the preparation. Examine with a microscope under low and high power.
4 Identify males and females (see *figure 9*).
5 Press gently on the coverslip, using the rubber end of a bulb pipette, to burst the body of a female and release the eggs.
6 Observe the eggs from a number of females and make a record of any stages in development that you can see. Use a graduated eyepiece and note the size of the eggs at various stages of development.
7 Locate an undivided egg and observe it over a period of time. Note the time of each division and the diameter of the developing egg.

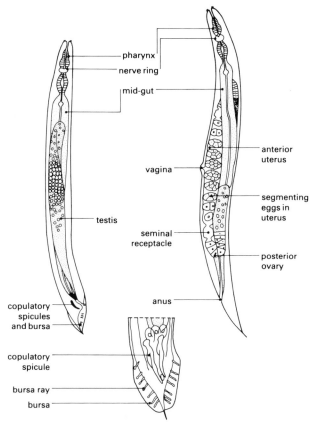

Figure 9

Diagrams of male (left) and female (right) *Rhabditis*, with an enlarged view of the posterior end of the male.

After Hinchliffe, J. R. 'Observation of early cleavage in animal development: a simple technique for obtaining the eggs of Rhabditis (*Nematoda*)', Journal of Biological Education, **7**(**6**), *1973, pp. 33–37.*

Questions

a *What pattern of cell division did you observe?*

b *What is the rate of cell division? Compare your results with others in the group and see if any pattern emerges.*

c *What happened to the egg size during cleavage? Explain the significance of your observations.*

d *In the light of your answer to c, what happens to the ratio of nuclear to cytoplasmic volume during cleavage? Why may this ratio be important?*

e *Consider the genetic information that is contained within each cell of the developing embryo. Would you expect it to be the same? Give reasons for your answers. What developmental problems does this raise?*

Early development in the flowering plant

Before you start the next investigation you must review the process of seed formation in flowering plants. The basic information is readily obtainable from a variety of texts. You will find that forming a seed is a complex business, involving a process of double fertilization resulting in the formation of a diploid, $2n$, zygote, and a triploid, $3n$, endosperm. The endosperm forms a nutritive tissue surrounding the developing embryo, providing it with food. In the mature seed of many angiosperms the endosperm remains as a food store for the germinating seedling. In other seeds, such as broad bean, *Vicia faba*, the food store is found in swollen embryonic leaves, the cotyledons, and not in the endosperm at all.

INVESTIGATION
23B Embryonic development in shepherd's purse, *Capsella bursa-pastoris*

(*Study guide* 23.2 'Early development'.)

Shepherd's purse is a small ephemeral plant, often found on urban wasteland and it is a common garden weed. The fruit, a capsule, is in the shape of a purse, or pouch, as was used by shepherds long ago – hence its common name. The species produces a number of generations during a year. We use this particular species for that reason, because it is almost always possible to find plants with developing embryos.

Procedure
1 Examine shepherd's purse plants. Find a plant which has small white flowers at the apex of the flowering shoot and developing capsules further down.
2 Remove a capsule. Place it in a solid watch-glass and, use a low power stereo binocular microscope and mounted needles, to dissect out the ovules. It is best to start with a larger, more mature capsule and then work up the shoot to smaller less developed ones.
3 Place an ovule in 5 per cent glycerine on a microscope slide. Add a coverslip and tap it with the handle of a mounted needle until the ovule bursts. It may help to leave the ovules in 5 per cent potassium (or sodium) hydroxide for a few minutes before placing them on a slide and squashing them.

4 Examine the burst ovule for the embryo, which is often found almost intact.

5 Use the above procedure to select ovules from a range of capsules of 1, 2, 3, 4, and 5 mm in length; examine them under low and high power.

6 If you do not get all five sizes, pool your results with other members of the group. Sketch the developing embryos to illustrate the changes as they develop.

Questions

a *In what ways is the process of cell division different from that found in* **Rhabditis***?*

b *At what stage were any future organs first discernible? Which organs were you able to identify?*

c *In the life cycle of many flowering plants the seed forms a dormant stage. What advantages does this have?*

INVESTIGATION
23C Morphogenesis in amphibians

(*Study guide* 23.3 'Development of the body plan'.)

After the initial process of cleavage has produced a mass of cells from the zygote, the cells arrange themselves into distinct regions. This cell proliferation and movement are the basis of morphogenesis. Early embryologists followed these cell movements by staining parts of the dividing embryo with a vital stain. This is a stain which will not affect the activities of the cell in any way. Cells from these marked areas could then be followed through subsequent development. As a result of such work it has been possible to construct 'fate maps' which show the structures that parts of the uncleaved and undifferentiated egg will eventually represent in the later embryo (*figure 10*).

Amphibians, especially the frog, were particular favourites of embryologists in the study of morphogenesis. Such techniques as vital staining and the tracing of cell movements are very exacting and time-consuming, and it is not really practical to carry them out in a school laboratory. However, it is possible to simulate the events using polystyrene spheres.

In this investigation you will construct three-dimensional models, based on the series of two-dimensional fate maps (*figures 11–15*), to show the movement of cell masses during the early development of a *Xenopus* embryo.

In cleavage, a mass of cells is formed. As this process continues a cavity,

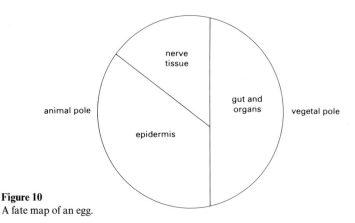

Figure 10
A fate map of an egg.

the blastocoel, forms inside the ball of cells. At this point the embryo is called a blastula. In the blastula the cells are arranged in two layers surrounding the blastocoel. *Figure 11* shows the fate map of the first, outermost cell layer. *Figure 12* shows the fate map of the second, inner layer, as we would see it if the outer layer were removed.

Following the blastula stage, gastrulation occurs. This is a process by which, through cell movements, a rearrangement of cell layers occurs and a gut is formed. The start (*figure 13a, b*) of gastrulation is marked by the inward migration of cells at the dorsal point of the prospective yolk plug – the region where the white yolky cells can be seen (*figure 13c*). This migration of cells forms a groove on the surface of the embryo – the blastopore groove (*figure 13d*). As the inward migration of cells continues the groove extends to form a complete circle of intucking (*figure 14*). The migrating cells move over the internal surface of the blastocoel, which, as a result, almost disappears. The sheet of surface cells spreads towards the posterior end and the yolk plug decreases in size as a new cavity, a primitive gut or archenteron, forms. *Figure 15* shows the cell regions at the completion of gastrulation.

Further cell movements result in an elongated embryo called a neurula, as shown in *figure 16*.

Procedure

1 Construct a model blastula by taking a polystyrene sphere and drawing onto its surface with felt pens the prospective fates of areas of the outer region. Use *figure 11* as a guide.

2 On a sheet of paper, draw the dorsal surface view of the blastula, with the animal pole to the left of your drawing.

3 Take a second sphere and cut it in half. Then, using *figures 11* and *12*, draw onto the cut surface of one hemisphere the cell layers and fates of different regions that would be seen in a vertical section through the

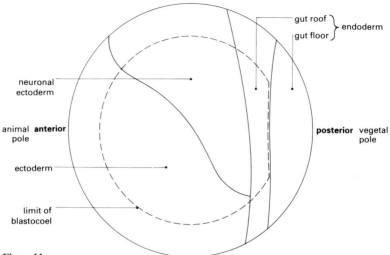

Figure 11

Fate map of the outer cell layer of the frog blastula.

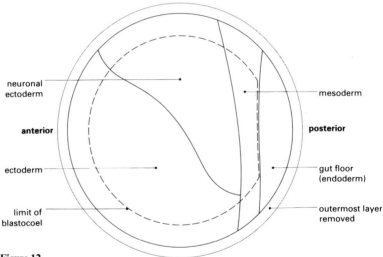

Figure 12

Fate map of the inner cell layer of the frog blastula.

middle of the blastula and cut in the plane of the paper in *figures 11* and *12*.

4   On a sheet of paper, draw the vertical section of the blastula.
5   Complete the right half-blastula model by drawing onto its curved surface the fate areas of the outer region.
6   Draw onto the cut surface of the remaining hemisphere, the cell layers and fate areas of a vertical section cut through the middle of the

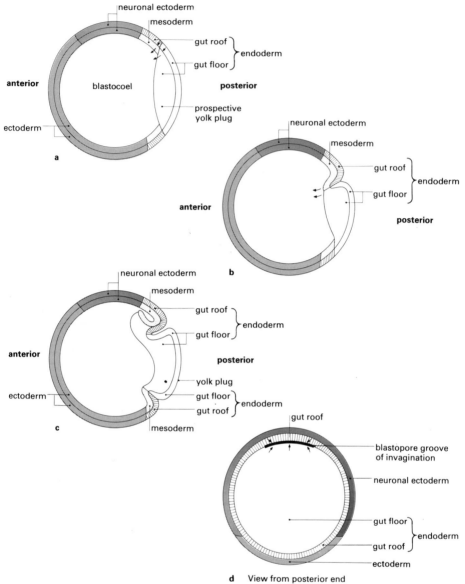

Figure 13
Fate maps of the frog embryo showing:
a and **b** The beginning of gastrulation.
c and **d** Formation of the yolk plug and the blastopore groove of invagination.

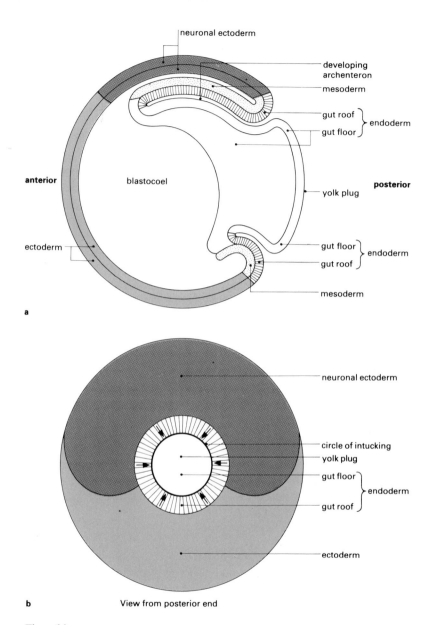

Figure 14
Fate maps of the frog embryo showing the formation of the gastrula and the inward migration of cells to form a circle of intucking.

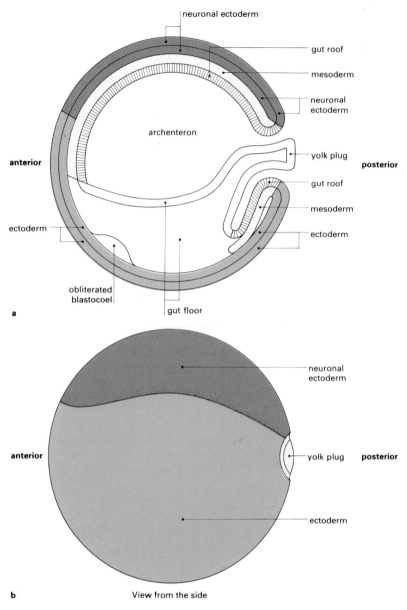

Figure 15
Fate maps showing the cell regions of the frog embryo at the completion of gastrulation.

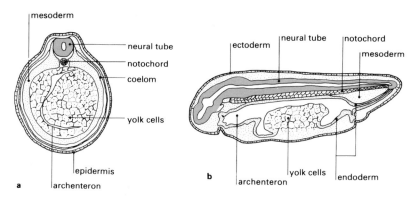

Figure 16
a A transverse section and **b** a longitudinal section of a neurula.
a *After Bodemer, C. W.,* Modern embryology, *Holt, Rinehart and Winston, 1968.*
b *After Cohen, J.* Living embryos, *2nd edition, Pergamon Press, 1967.*

blastula at right angles to the plane of the paper (*figures 11* and *12*) as if viewed from the posterior end.

7 On a sheet of paper, draw the vertical section as seen from the posterior end.

8 Complete the anterior half-blastula model by drawing onto its curved surface the fates of different areas of the outer region.

9 Using *figure 15*, draw on a sheet of paper:
(a) a dorsal view of the gastrula
(b) a vertical section through the middle at right angles to the plane of the paper and viewed from the posterior end.

Questions

a *What happens to the prospective mesoderm cells of the blastula during gastrulation?*

b *What are the developmental fates of the cell layers in the gastrula?*

c *Compare the structure of the gastrula with that of the neurula (figure 16). What changes in form occur during neurulation?*

INVESTIGATION
23D Growth and development of the fore-limb of mice

(*Study guide* 23.5 'Growth' and 23.8 'Pattern formation in the development of limbs'.)

Although mice embryos are highly organized and specialized by nine days after fertilization, there are still structures that have to undergo

10 day   11 day

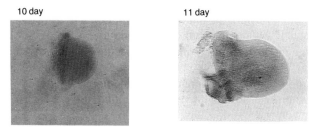

Figure 17
Early limb buds in the mouse (× 18)
Photographs (figures 17–21), Dr Cheryll Tickle, Department of Anatomy and Biology as Applied to Medicine, The Middlesex Hospital Medical School, London.

their developmental sequence. At about ten days the fore-limb buds begin to be recognizable (*figure 17*).

By differential staining for cartilage in a range of different aged embryos the growth and differentiation of the fore-limbs can be followed. This investigation will involve you in taking measurements of the limb elements from the photographs shown in *figures 19* to *21*. *Figure 18* will help you to identify the different elements of the limb.

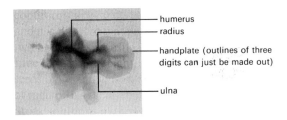

humerus
radius
handplate (outlines of three digits can just be made out)
ulna

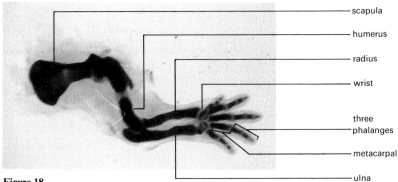

scapula
humerus
radius
wrist
three phalanges
metacarpal
ulna

Figure 18
The elements of mouse fore-limbs (× 9).

12–12$\frac{1}{2}$ day

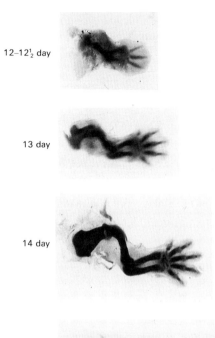

13 day

14 day

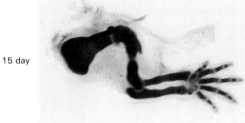

15 day

16 day

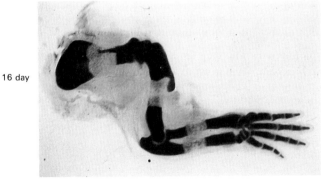

Figure 19
Mouse fore-limb development series 1 (× 9).

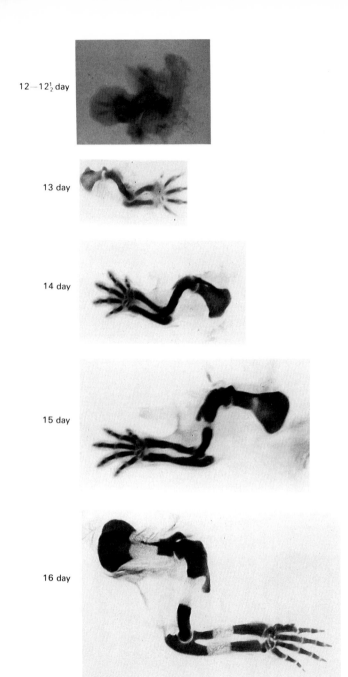

$12 - 12\frac{1}{2}$ day

13 day

14 day

15 day

16 day

Figure 20
Mouse fore-limb development series 2 (× 9).

12—12$\frac{1}{2}$ day

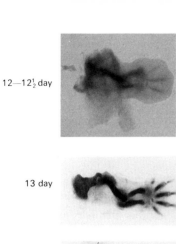

13 day

14 day

15 day

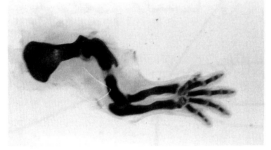

16 day

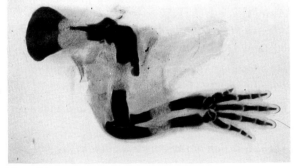

Figure 21
Mouse fore-limb development series 3 (× 9).

Procedure

1   With a pair of vernier callipers, measure the length in millimetres, of the humerus, radius, wrist, and middle digit of each limb in the photographs of series 1 (*figure 19*).

2   You will need to look closely at each photograph and decide the precise points between which to take measurements for each limb component.

3   Repeat steps **1** and **2** for the photographs in series 2 and 3 (*figures 20* and *21*).

4   Calculate the average length of each limb component and the average total limb length for 12-, 13-, 14-, 15-, and 16-day old mice.

5   Collect the average values from step **4** from other members of the group and use these to calculate the average for the group. Present the data in the way you think is most appropriate.

6   Use your group averages to draw a growth curve, length against time, for each limb component and for the limb as a whole.

7   Draw growth rate curves by plotting increase in length against time.

8   Draw percentage growth curves by plotting increase in length as a percentage of the previous day's length.

Questions

a   *What can you deduce from the three types of growth curve that you have drawn?*

b   *What is the pattern of growth during mouse fore-limb development?*

c   *Examine the growth rate curve from procedure 7. What changes in proportion of the limb elements have occurred during development?*

d   *What two processes could explain the drop in growth rate of the wrist and middle digit during day 13?*

e   *Using the percentage growth curves from procedure 8, estimate the age of the mouse at which the fore-limb would stop growing. Explain how you arrived at your answer and what assumptions, if any, you made.*

f   *Why might percentage growth curves be more useful to developmental biologists than absolute growth curves?*

g   *What is the nature of the light unstained regions that appear in the photographs of 15 and 16 day-old limbs?*

h   *What does the location of these lighter regions suggest about the growth and differentiation of cartilage during long bone development?*

INVESTIGATION

23E Growing new plants from old

(*Study guide* 23.9 'The pattern of plant and animal development compared'.)

During growth and development, the cells of living things differentiate, some becoming organs with definite functions. In the mature organism, differentiated cells exhibit certain powers of regeneration, usually as the result of wounding. That suggests that their developmental fate is not rigidly fixed.

Some animal groups have considerable powers of regeneration. A planarian, for example, if cut in half, can regenerate a new anterior and posterior end. Earthworms also have powers of regeneration, although not nearly as great as popularly thought. Newts can regenerate a limb and many lizards a tail, although the latter is usually not as long and graceful as the original. Humans too are capable of limited regeneration – the healing of open wounds is an example. All these processes involve redifferentiation.

In contrast with animal groups, plants show this ability to regenerate much more dramatically. In this investigation you will use a simple method of tissue culture which will enable you to grow a new plant from a small part of a mature one.

Procedure

1 Make up a sterilizing solution by mixing 1 cm³ of domestic bleach with 9 cm³ of water. Place the solution in a McCartney bottle and replace the cap.

2 Sterilize your working surface by wiping it with a towel soaked in disinfectant. Rinse with water and dry with a clean towel.

3 Cut off one medium-sized floret from the curd of a cauliflower. Score the surface of the floret, with a scalpel, into squares with size of about 3 mm.

4 By making a horizontal cut 3 mm deep, cut off cubes of sides 3 mm, about the size of a match head. These are the explants.

5 Place the explants in the bottle of sterilizing fluid. Should any fall onto the working surface, ignore them and cut off fresh explants.

6 Screw the cap onto the bottle, shake the contents vigorously for one minute, and allow to stand for ten minutes.

7 Using a sterile inoculating loop, transfer the explants from the sterilizing solution to a bottle containing sterile water. Replace the cap and shake the contents for a minute.

8 Then, using a sterile inoculating loop, transfer the explants to a second bottle of sterile water and treat as before. Repeat with a third bottle.

Figure 22
A cauliflower plantlet.
Photograph, National Vegetable Research Station, Wellesbourne, Warwickshire.

9 Finally, using sterile technique, take a bottle of culture medium and, with a sterile loop transfer the explants from the water onto the surface of the medium. Replace the cap of the bottle of medium, but screw it up only until it is finger-tight. Repeat with one or two more bottles of culture medium.

10 Keep the culture bottles at about 28 °C and under constant illumination for 2–3 weeks.

11 Sow 10 to 20 cauliflower seeds to a depth of about 5 mm in a tray of moist seed compost.

12 Use a hand lens or a low power stereo binocular microscope to observe the explants and the seed tray every two or three days. Record any changes. Ensure that the seed containers do not dry out and do not open the culture bottles.

Questions

a *What would the cells of the curd normally have developed into? From your observations of the explants, what changes in structure and function did these cells undergo?*

b *What conclusions can you draw concerning the ability of the explant cells to develop their genetic potential?*

c *Why does the curd of a cauliflower not show the sort of changes that you found in the explants?*

d *How many new cauliflower plants could be grown by taking explants from one cauliflower curd? Explain how you calculated your answer.*

e *Plants grown by such a culture technique are called clones. What is the meaning of the term? Why may the process be useful in horticulture and agriculture?*

f *Compare the two methods of propagation used in this investigation. What are the advantages and disadvantages of each?*

g *Review the different methods of tissue culture. What are the potential applications for these techniques?*

CHAPTER 24

CONTROL AND INTEGRATION THROUGH THE INTERNAL ENVIRONMENT

INVESTIGATION

24A **The microscopic structure of endocrine glands**

(*Study guide* 24.1 'Endocrine communication and control in animals' and 24.2 'Hormone synthesis and release'.)

Endocrine glands are where the synthesis and secretion of hormones take place. These glands do not have ducts and the hormones, which act on specific target organs, are transported throughout the body in the blood.

Question

a ***What characteristic features would you expect endocrine glands to have?***

The three endocrine glands which it is suggested that you examine under the microscope are the pancreas, the adrenal medulla, and the thyroid gland.

The pancreas (see *Study guide* 24.2, 'Hormone synthesis and release'). The pancreas has two roles. It secretes a fluid rich in digestive enzymes into the duodenum via the pancreatic duct, which is its exocrine function. It also secretes the hormones insulin and glucagon, which is its endocrine function.

The adrenal medulla (see *Study guide* 24.2, 'Hormone synthesis and release'). The adrenal medulla secretes the catecholamine hormones, noradrenaline and adrenaline, under the control of the sympathetic nervous system.

The thyroid gland (see *Study guide* 24.2, 'Hormone synthesis and release'). The thyroid gland secretes the iodine-containing hormones tri-iodothyronine (T_3) and thyroxine (T_4). The thyroid gland is unique among human endocrine glands in that large amounts of the hormone are stored in an inactive form – thyroglobulin – in follicles. (Other endocrine glands may store small amounts of hormones inside their cells.)

Procedure

1   Examine a section of pancreas under the low-power magnification of a microscope. Identify the enzyme-secreting cells and the islets of Langerhans with the help of *figure 23*.

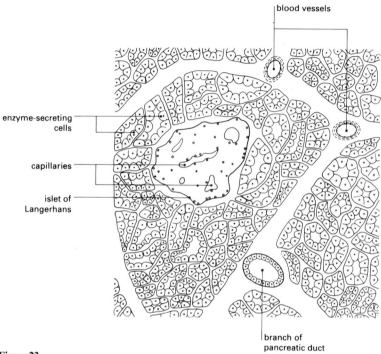

blood vessels

enzyme-secreting cells

capillaries

islet of Langerhans

branch of pancreatic duct

Figure 23
A section of the pancreas.
After Freeman, W. H. and Bracegirdle, B., An atlas of histology, *2nd edition, Heinemann Educational Books, 1967.*

2   Now examine an islet of Langerhans (the endocrine part of the pancreas) under high power. Make a drawing to show the arrangement of the hormone-secreting cells and the capillaries.

   In preparations which have been suitably stained two kinds of secretory cells, called α- and β-cells, may be distinguished. The α-cells secrete glucagon and the β-cells secrete insulin.

3   Either by studying your own preparation or from the photomicrograph in *figure 24*, note any differences you can see between the two types of cell.

4   Examine a section of the medulla (middle part) of the adrenal gland under low and high power (*figure 25*). Record the arrangement of the secretory cells and blood capillaries.

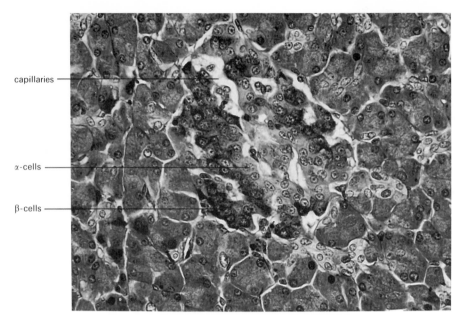

capillaries

α-cells

β-cells

Figure 24
A photomicrograph of a section through an islet of Langerhans (× 325). The pale-stained cells are α-cells and the dark-stained ones β-cells.
Photograph, Biophoto Associates.

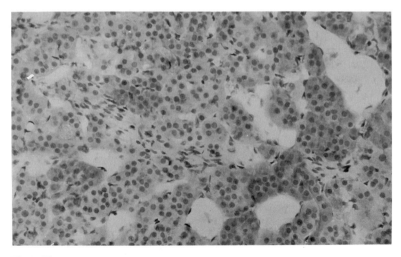

Figure 25
The adrenal medulla (× 80). The tissue is composed of closely packed clumps of secretory cells supported by a fine reticular network containing numerous wide capillaries.
Photograph, Biophoto Associates.

5 Examine a section of thyroid gland under low and high power (*figures 26* and *27*). Identify the secretory cells, the thyroglobulin stored in the follicles, and blood capillaries.

follicle

basement membrane (very thin)

endocrine secretion, thyroglobulin, accumulating in follicle

simple cuboidal epithelium

blood vessel

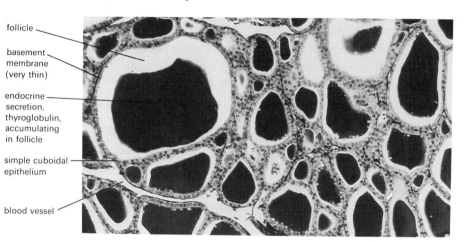

Figure 26
A section through the thyroid gland, showing follicles filled with thyroglobulin ($\times 290$).
Photograph, Biophoto Associates.

Golgi complex

mitochondrion

rough endoplasmic reticulum

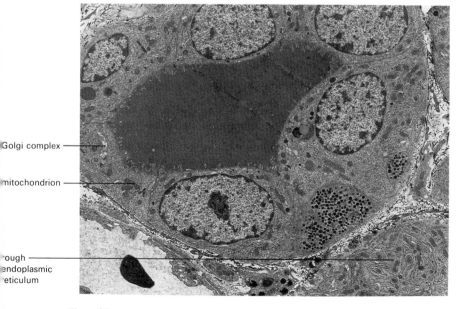

Figure 27
An electronmicrograph of a thyroid follicle ($\times 4600$).
Photograph, Biophoto Associates.

Questions

b *Are the characteristic features you predicted for endocrine glands in your answer to question **a** present in the glands you have examined?*

c *In what ways are the glands you studied dissimilar?*

d *In the electronmicrograph shown in* figure 27 *the rough endoplasmic reticulum is a prominent feature. You can also see a Golgi complex and mitochondria. How does the presence of these structures relate to the hormone-producing function of the cell?*

INVESTIGATION
24B The effect of IAA on the growth of coleoptiles and radicles

(*Study guide* 24.7 'Communication and control in plant growth'.)

Several chemical compounds extracted from plant tissue have been shown to affect the responses of plants to stimuli. One of the commonest has been identified as indole-3-ethanoic (acetic) acid (IAA).

Procedure

1 Prepare five Petri dishes, each containing one of a range of IAA and sucrose solutions, as follows:
(a) $10\,cm^3$ stock IAA solution ($10^{-2}\,mol\,dm^{-3}$ IAA in sucrose buffer)
(b) $1\,cm^3$ solution (a) $+\,99\,cm^3$ pure sucrose buffer (use $10\,cm^3$)
(c) $1\,cm^3$ solution (b) $+\,99\,cm^3$ pure sucrose buffer (use $10\,cm^3$)
(d) $1\,cm^3$ solution (c) $+\,99\,cm^3$ pure sucrose buffer (use $10\,cm^3$)
(e) $10\,cm^3$ pure sucrose buffer
The concentrations of IAA will be 10^{-2}, 10^{-4}, 10^{-6}, 10^{-8}, and $0\,mol\,dm^{-3}$ respectively.

2 Choose 25 coleoptiles of about 18–20 mm length. Line them up on a wet microscope slide, five at a time, with their tips against one edge. Cut 10 mm sections, starting 3 mm back from the tip, as shown in *figure 28*. Store the sections in distilled water until they are needed. Repeat with radicles.

3 Place five coleoptile and five radicle sections in each dish and incubate, in the dark, for 24–48 hours at 25 °C.

4 Measure the length of each section to the nearest 0.5 mm using a ruler and hand lens or vernier callipers. Subtract 10 from each reading to find the increase in length from the original (10 mm).

5 Pool the class results and calculate the mean and standard deviation for each treatment (there are five treatments for coleoptiles and five for radicles).

6 Plot graphs of mean increase in length of coleoptiles and radicles

a

two razor blades held 10 mm apart

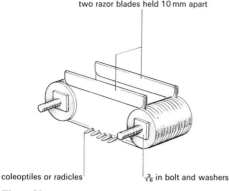

coleoptiles or radicles $\frac{3}{16}$ in bolt and washers

b   a piece of metal exactly 10 mm wide
placed over 5 coleoptiles

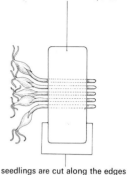

seedlings are cut along the edges

Figure 28
Ways of cutting 10 mm sections.

against $\log_{10}$ of IAA concentration. Mark the standard deviation for each treatment on the graphs.

7 Calculate the standard error of the difference between the control (treatment (e)) and each of treatments (a) to (d), using the formula

$$\text{standard error of the difference} = \sqrt{\frac{S_1^2}{N_1} + \frac{S_5^2}{N_5}}$$

where S_1 and S_5 are the standard deviations from treatments (a) and (e) respectively, and N_1 and N_5 are the number of results obtained in each treatment. If the difference between the means of the two treatments is greater than or equal to twice the standard error of the difference then it may be concluded that the difference is not due to chance, but to a genuine biological response to the treatments.

Questions

a   *From the graphical analysis, does IAA appear to affect the coleoptile and radicle sections? If so, is the effect roughly proportional to the concentration of IAA?*

b   *In what ways do different concentrations of IAA affect the coleoptile and radicle sections?*

c   *From the statistical analysis, which concentrations of IAA appear to have a significant effect on the sections?*

d   *In what ways do the results obtained with IAA aid the understanding of tropic responses?*

e   *Why do you discard the 3 mm at the tip of the section?*

INVESTIGATION

24C Stimulation of amylase production in germinating barley grains

(*Study guide* 24.7 'Communication and control in plant growth'.)

The bulk of a barley grain consists of a store of reserve food – the endosperm. The embryo occupies only a very small proportion of the total volume and it is situated at the more pointed end on the opposite side from the groove. It can only be seen when the husks have been removed. During germination the food reserves must be made soluble for transport to the embryo to support continued growth of the seedling. This investigation examines how this is brought about. (*Note:* It is essential that a good sterile technique is used throughout this investigation.)

Procedure

1 Dehusk and cut the barley grains across the line AA (*figure 29*) so that one half contains the embryo and the other only food reserve. Keep the two halves separate.

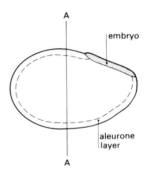

Figure 29
A longitudinal section through a barley grain.

2 Tie the halves separately in a bag made of muslin and sterilize them by putting them in 3 per cent sodium hypochlorite solution for 5 minutes. Wash thoroughly, but quickly, through five changes of sterile water until there is no further smell of chlorine.

3 Take two sterile Petri dishes containing starch agar. Place five of the embryo-containing halves, cut face downwards, on the agar in one and five non-embryo halves in the other.

4 Then take a sterile Petri dish containing starch agar to which 0.1 per cent gibberellic acid has been added to give a final concentration of 10 p.p.m. Place five non-embryo halves cut face downwards on the agar.

5 Incubate the plates at 25–30 °C for about 48 hours.

6 Remove the plates from the incubator and pour a solution of iodine dissolved in potassium iodide all over the surface of the agar in each dish. Record your observations.

(*Note:* Infection by various micro-organisms is fairly common if the sterile technique is not carefully applied and these may also produce an amylase.)

Questions

a *Under which conditions is a starch-degrading enzyme (amylase) produced?*

b *Suggest a hypothesis for the control of amylase production in germinating barley grains.*

To investigate this process further carry out the following experiment in which two halves of the grain are separated by a selectively permeable membrane. The protein molecules of the amylase are too large to pass through it while the smaller gibberellic acid molecules should get through.

7 Take a sterile Petri dish containing starch agar into which a 'boat' made of dialysis tubing containing plain agar has been placed.

8 Inside the boat place five embryo-containing halves of the barley grains, cut face downwards. Alternatively (or as an additional

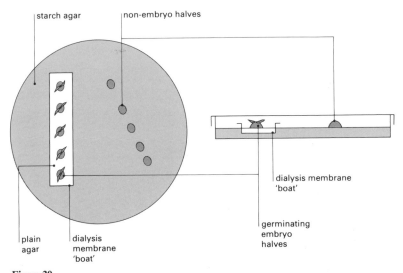

Figure 30
The arrangement for separating the two halves of a barley grain by a selectively permeable membrane.

experiment), you could put agar containing gibberellic acid inside the boat and no embryo-containing halves of barley grains.

9   Outside the boat on the starch agar place five non-embryo halves, to act as 'detectors' for any diffusible stimulus to amylase activity (see *figure 30*).

10   Incubate the plate for about 48 hours, then irrigate with a solution of iodine dissolved in potassium iodide. Record your observations.

c   *What can you deduce from these results?*

d   *What part of the barley grain would you suggest produces the amylase? Look carefully at the base of the grains that have produced amylase to see if their appearance gives a clue.*

e   *What is the significance of your findings in seed germination and the control of dormancy?*

INVESTIGATION
24D   The effect of plant hormones on seed germination

(*Study guide* 24.7 'Communication and control in plant growth'.)

Seeds will sometimes fail to germinate even when all the external conditions appear to be favourable. This condition is called dormancy and it is thought that plant hormones may have a role in the process. The two hormones investigated are abscisic acid (ABA) and kinetin (KIN).

1   Line the bases of eight Petri dishes with filter paper and count out 50 seeds of a light-sensitive variety of lettuce such as Dandie into the lid of each dish.

2   Stock solutions of the two plant hormones are provided:

ABA $0.002\,\mathrm{g\,dm^{-3}}$ (2 p.p.m.)
KIN $0.02\,\mathrm{g\,dm^{-3}}$ (20 p.p.m.)
These are diluted 1:1 with water or the other hormone before use so that the final concentrations are:
ABA $0.001\,\mathrm{g\,dm^{-3}}$ (1 p.p.m.)
KIN $0.01\,\mathrm{g\,dm^{-3}}$ (10 p.p.m.)
A convenient way of doing this is to use the stock solutions and add $5\,\mathrm{cm^3}$ of the appropriate hormone(s) or water to the filter paper in the Petri dishes as follows:
(a) $2.5\,\mathrm{cm^3}$ ABA + $2.5\,\mathrm{cm^3}$ water
(b) $2.5\,\mathrm{cm^3}$ KIN + $2.5\,\mathrm{cm^3}$ water
(c) $2.5\,\mathrm{cm^3}$ KIN + $2.5\,\mathrm{cm^3}$ ABA
(d) $5.0\,\mathrm{cm^3}$ water
Prepare two sets, one for the light and one for the dark. Label them.

3 Remove any air bubbles that may form as the filter papers expand on wetting, then scatter the 50 seeds evenly over the surface.

4 Place one set in the light and the other in the dark at 25 °C.

5 Examine and record percentage germination after 2–4 days.

Questions

a *What effect does each hormone have on the germination of lettuce seeds?*

b *What is the effect when both hormones are present?*

c *What evidence do you have from your results that these plant hormones have a role in the control of dormancy in plants?*

CHAPTER 25

DEVELOPMENT AND THE EXTERNAL ENVIRONMENT

INVESTIGATION
25A **The effect of temperature on root growth**

(*Study guide* 25.2 'The external environment in relation to growth and development'.)

Procedure

Before starting this investigation decide on two temperatures at which to take measurements. The temperatures should be in the range 15–35 °C, and 10 °C apart.

Then assemble the boiling tube and beaker as shown in *figure 31*.

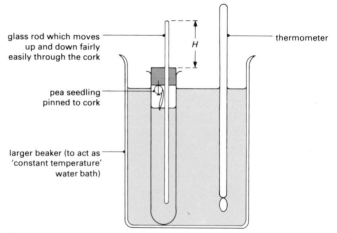

glass rod which moves up and down fairly easily through the cork

thermometer

H

pea seedling pinned to cork

larger beaker (to act as 'constant temperature' water bath)

Figure 31
Apparatus for measuring the effect of temperature on root growth.

1 Select a pea seedling with a root about 2 or 3 cm long and pin it to the cork, with the glass rod in position.

2 Mix hot and cold water from the taps in the large beaker until the mixture is at the lower of the two required temperatures. (*Note:* It may be necessary to stir a little more hot or cold water into the beaker in order to maintain the desired temperature during the course of the experiment.)

3 Use the water from the beaker to fill the boiling tube until the water just covers the apical 4–6 mm of root tip when the glass rod is almost touching the bottom, leaving at least 2 or 3 cm protruding beyond the cork.

4 Place the assembled boiling tube in the beaker and leave for five minutes for the root temperature to equilibrate with the water temperature.

5 To make a measurement, slowly raise the glass rod through the cork until the meniscus between the water in the boiling tube and the root tip just breaks.

6 Now very gently lower the rod until the meniscus just 'jumps' back on to the root apex.

7 Measure the length of the glass rod (H) protruding above the top of the cork to the nearest millimetre.

8 Take two more similar readings in quick succession and note the time at the second reading. Use the average of the three values of H. Take five sets of readings in this way at ten-minute intervals.

9 After each set of readings lower the glass rod so that the water covers the bottom 5 mm of the root tip until the next reading. This makes sure that root and water are at the designated temperature and prevents water stress in the root.

10 Plot a graph of the mean value of H against time. The slope of the line represents the rate of root growth at that temperature.

11 Now obtain a second series of measurements in the same way at a temperature 10 °C higher than the first, again filling the tube from the beaker after discarding the cooler water.

12 Allow five minutes for the root temperature to equilibrate with the water temperature before starting the new set of measurements.

13 Plot these measurements on the same graph as the earlier data.

14 Calculate the rate of growth at each temperature (change in H per unit time) and determine the temperature coefficient (Q_{10}) for the root growth process as follows:

$$Q_{10} = \frac{\text{Rate at } (t + 10)\,°C}{\text{Rate at } t\,°C}$$

Questions

a *What information do the slopes of the graphs give you about the rate of root growth?*

b *What effect would you expect further increases of 10 °C to have on the rate of root growth? Give your reasons.*

INVESTIGATION
25B Effects of light on the germination of lettuce seeds

(*Study guide* 25.3 'Light and plant growth'.)

The amount of time that light lasts, its intensity and quality (wavelength), and the cycle of light and dark have a profound effect on the growth and development of plants.

It is easier and quicker to detect the effect of light on the early stages of plant development than on later ones. For this reason the following investigation makes use of seeds. It is based on the results of some classic experiments using lettuce seeds, mostly performed in the 1950s. Light is only one of the environmental factors that can affect seed germination. Only a few varieties of lettuce seeds are sensitive to light during germination; they tend to be early-forcing varieties which are used for glasshouse production of lettuce in winter.

In order to investigate the role of light in germination, we need to have some knowledge of the nature of light itself. White light is only a

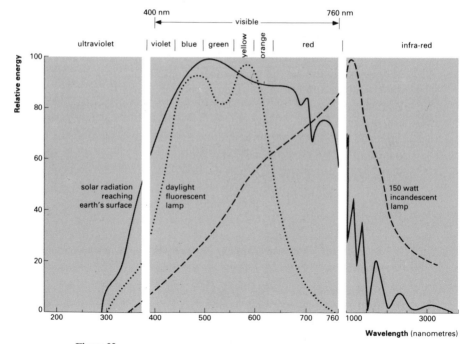

Figure 32
Sources of illumination and distribution of wavelengths in nanometres (10^{-9} m). Note the break and change of scale in the wavelength axis.
Based on Biological science: interaction of experiments and ideas, *(1966) p. 180; by kind permission of the Biological Sciences Curriculum Study.*

tiny part of a wide band of electromagnetic radiation consisting of waves ranging in length from billionths of a millimetre to thousands of metres. *Figure 32* gives an idea of the position occupied by visible light, which is bounded on one side by violet and the other by red, the shortest and the longest wavelengths detected by the human eye. Other eyes may differ in their range; for example, insects can recognize ultra-violet patterns which we cannot see.

If an object in sunlight reflects all visible wavelengths, it appears white; if it reflects none it is black. Those wavelengths which are not reflected are either absorbed or transmitted. A piece of clear glass transmits most of the light falling on it while coloured glass transmits its own colour but absorbs and reflects the others. When a transparent object appears green it transmits green light, but when an opaque object (such as a leaf) appears green it reflects green light, most of the remaining colours being absorbed.

Figure 32 shows the distribution of wavelengths in the light emitted by different sources, given as a percentage. It will be seen that when attempting to reproduce sunlight in the laboratory it is best to use a combination of fluorescent and incandescent lamps.

Procedure A: Effect of light on germination
You need to set up test conditions for germination to find out whether light promotes, inhibits, or has no effect on germination. At the same time you can see if a short period of light has the same effect as a long one. To see if temperature is important you can use two different temperatures.

Table 1 indicates one way of achieving all of these conditions.

	Light		Dark
	30 s	300 s	continuous
Room temperature			
25 °C			

Table 1

1   Take two varieties of lettuce seed, such as Dandie (or Kloek) and Great Lakes (or Webbs Wonderful).
2   Set up the number of Petri dishes you need as follows. Line each Petri dish base with filter paper. Count out 50 of the required seeds into the lid of each Petri dish. Add about 5 cm^3 of water to the filter paper and remove any air bubbles which may form. Then add the 50 seeds spread

evenly over the surface. Make sure the dishes are labelled, preferably on their bases (lids can become interchanged when scoring germination).

3   Leave the seeds to soak for one hour *in complete darkness* at room temperature. The seeds are not sensitive to light until after they have taken in water.

4   After this time put the seeds that are to receive light under a 40 watt lamp 30–60 cm above the bench for the required length of time. The lamp must be suitably shielded. Keep one set at room temperature and the other at a controlled temperature of 25 °C.

5   Check there is enough water in the Petri dishes and put the seeds into the dark, keeping one set at room temperature and the other at a controlled temperature of 25 °C. The seeds left in continuous dark must be in complete darkness from the start of the experiment.

6   Examine after two or three days and record the percentage germination in all the conditions. Count only those which have a radicle 1 mm long, then twice this number equals the percentage germination. (It is easy to confuse a shed testa with an ungerminated seed.)

Questions

a   *What is the effect of light on the germination of the two varieties of lettuce?*

b   *Does the length of time for which the seeds are exposed to light have any effect?*

c   *Is temperature a critical factor?*

d   *In photosynthesis light is used as a major source of energy. Do you think this is likely to be the case in germination, or from your results can you suggest another role for it?*

Procedure B: Effect of different qualities of light on germination
To investigate this you must be able to control the wavelengths of light used. You can do this by removing (filtering) from natural or artificial light those wavelengths you do not require, so transmitting only those you need for the experiment. *Figure 33* shows the transmission data for a number of cinemoid filters. Far-red light can be obtained by putting a red and blue filter together. This is because this blue filter has a second band of transmission starting around 720 nm, which we cannot see. Thus the blue filter stops the red transmission, the red filter stops the blue transmission, but they both transmit beyond 720 nm in the far-red region.

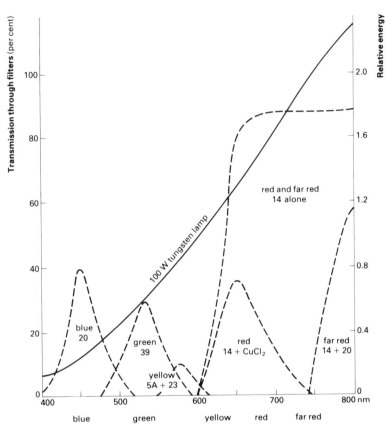

Figure 33

Transmission characteristics of various Cinemoid filters.
From Hannay, J. W. 'Light and seed germination – an experimental approach to
photobiology', Journal of Biological Education, **1**, *1967, pp. 65–73.*

1　Sow the required number of sets of a light-sensitive variety of seed and
　let them imbibe water as in Procedure A.
2　Using different coloured filters expose the seeds to different
　wavelengths of light for 300 s (providing this time was satisfactory in
　Procedure A) and then return them to darkness at 25 °C.
3　Set up seeds with 300 s white light and continuous darkness as
　controls.
4　Examine after two or three days and record the percentage
　germination under each wavelength of light.

Procedure C: Effect of alternating the promoting and inhibiting wave-
lengths of light
The results from Procedure B will probably have indicated that one filter

seemed to inhibit germination while another caused a germination rate as high as that of white light. You can confirm these effects and gain some further knowledge of the system if you expose seeds to these two filters alternately.

1   Let the required number of sets of seeds imbibe water as in the previous procedures.

2   Expose the seeds to the two filters alternately but give twice as long under the filter which inhibits (i.e. 600 s). Alternate the filters as many times as you like, but make sure you have one set of seeds which was last exposed to one filter and one to the other. Note which was which, then return the seeds to darkness at 25 °C (be careful to exclude white light when alternating the filters).

3   After two or three days record the percentage germination for each set of seeds.

e   *How do different wavelengths of light affect germination? Which wavelengths appear to promote and which appear to inhibit germination? Are these effects reversible?*

f   *Why would fluorescent lamps be useless as the sole light source for these experiments?*

g   *From these experiments formulate a hypothesis involving a pigment to explain the mechanism concerned in influencing seed germination. What colour would you expect the pigment to be?*

h   *In view of the inhibiting effect of one wavelength of light, suggest a reason why light-sensitive seeds will germinate in white light, which contains all wavelengths.*

Procedure D: Use of a green leaf as a light filter

1   Take the two varieties of lettuce seeds as in Procedure A and allow as many sets as you need to imbibe water.

2   Take a large green leaf such as French or runner bean or sycamore and cut it so that it covers one half of each of the Petri dishes. Group the seeds into two sets of 25, making sure that those under the leaf are well covered by it (*figure 34*).

3   Expose the seeds to different wavelengths of light as in Procedure B but this time give prolonged (16 hours per day) or continuous illumination at room temperature and at 25 °C. Include controls in white light and the dark.

4   Examine after two or three days and record the percentage germination, noting any differences between the leaf-covered seeds and those exposed directly to the light.

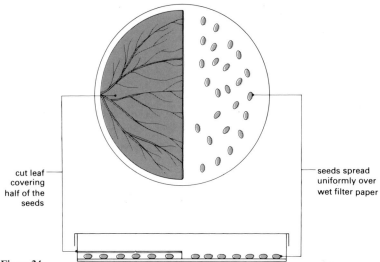

Figure 34
The arrangement for using a green leaf as a light filter.

i *What is the effect of a green leaf filter on the germination of the two varieties?*

j *Is temperature a critical factor for this effect?*

k *What is the ecological significance of the effect of a green leaf filter?*

Study item 25B.1

An experiment was set up to investigate the effect of light on the germination of the seeds of the plant *Lycopus europaeus* (gipsywort). Five batches of seeds were initially soaked for two hours in the dark at the temperatures given in table 2. Each batch was then exposed to a different cycle of temperature fluctuation. At the same time, within each batch, half of the seeds were exposed to alternating light and dark cycles and the other half were kept in continuous darkness. The treatments and results are shown in the table.

Batch treated	Temperature during preliminary soaking (°C)	Temperature cycles (°C)	Percentage germination: in cycles of dark/light	in continuous darkness
1	25	25/25	0	0
2	25	15/25	64	0
3	25	10/25	88	0
4	15	15/15	0	0
5	10	10/10	0	0

Table 2

Questions

a *Give two reasons why the seeds were soaked in the dark for two hours before the experiment.*

b *Suggest a possible reason why there is no result showing 100 per cent germination*

c *Using the table of results, give two conditions which are shown to be necessary for germination.*

d *No experiments were carried out in this case on the wavelength of light involved in germination. From your knowledge of other work on the effect of light wavelength on the physiology of germination, outline how you would test the hypothesis that the production of far*
☐ *red phytochrome (P_{735}) is necessary for germination.*

INVESTIGATION
25C Early environment and later behaviour of mice

(*Study guide* 25.4 'The study of ontogeny'.)

It has long been realized that in humans the early environment in which a child is brought up exerts a profound effect on behaviour and achievement later on.

The following investigation provides a means of determining the effect of one type of change in early environment and also a way of assessing later behaviour quantitatively in order to compare it with a control.

Procedure

1 To start the investigation you will require two litters of newly born mice. These must be either from the same inbred strain or from F_1 crosses. Keep the two litters in separate cages.

2 With one litter, remove each young mouse daily from the first day after birth and very gently pass it from hand to hand about forty times. Then replace the animal in the nest. Continue this treatment until the eyes open at about the twelfth day.

3 Rear the other litter normally. It is essential to disturb the animals as little as possible until after the twelfth day and it is best to rear them in isolation from other mice.

4 After the twelfth day rear both litters normally and wean them at about the twenty-first to twenty-fifth day.

5 Until you have tested the animals for their behaviour, keep the two groups separate. The activity of a mouse can be measured in an 'open field' apparatus (*figure 35*). This is basically an area surrounded by a

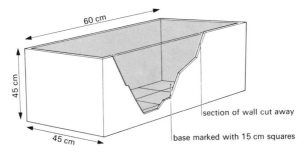

60 cm

45 cm

45 cm

section of wall cut away

base marked with 15 cm squares

Figure 35
Open field apparatus.

wall, the floor of which is marked out in squares so that the movement of the animal can be followed. By recording the amount of movement, you can measure the 'exploratory behaviour' of the animal. If you record the number of faecal pellets deposited in a given time, this will provide you with a measure of the 'emotionality' of the animal.

6 Set up the 'open field' and arrange a 40 watt lamp, suitably shielded, directly over the centre of the square, high enough to illuminate it as evenly as possible. It is best if the rest of the room is darkened, and it is important to reduce the noise level to a minimum during the test.

7 Place the mouse to be tested in one of the peripheral squares. When transferring the animal, pick it up by the tail about one third of the way from the body and carry it gently, but quickly.

8 Record the activity of the animal for three minutes, noting the amount of movement from one square to another and the number of faecal pellets deposited.

9 At the end of the period, return the animal to a cage, clean the surface of the apparatus, and repeat with another mouse.

10 You should repeat the test over a period of time, if possible at daily intervals. When you do so, select the animals in a random order. You can then record the mean activity scores for the two litters over the period. Apply a statistical test to find out if the differences are significant.

Questions

a *Why are inbred strains or F_1 litters used?*

b *Why are animals removed in a random order when repeating the 'open field' test?*

c *Why was it necessary to illuminate the arena as uniformly as possible and to reduce the noise level during testing?*

d *Apart from differences in the amount of movement of the animals, were any other aspects of their activity noticeable?*

e *What, if any, was the relationship between the 'exploratory behaviour' and 'emotionality' of the animals as measured in the investigation?*

f *If you found a significant difference in the behaviour of the two groups, how easy was it to assign an individual animal to one or other category?*

SUGGESTIONS FOR FURTHER READING

BRACEGIRDLE, B. and MILES, P. H., *An atlas of plant structure*, Volume 1. Heinemann, 1971.
(Angiosperm embryo development.)

BUTCHER, D. N. and INGRAM, D. S., Studies in Biology No. 65, *Plant tissue culture*. Edward Arnold, 1976.
(Details of the variety of techniques available.)

FREEMAN, W. H. and BRACEGIRDLE, B., *An atlas of histology*. Heinemann, 1967.
(Structure of mammalian gonads.)

FREEMAN, W. H. and BRACEGIRDLE, B., *An advanced atlas of histology*. Heinemann, 1976.
(Structure of mammalian gonads.)

INGOLD, C. T., Studies in Biology No. 88, *The biology of* Mucor *and its allies*. Edward Arnold, 1978.
(Life cycle of *Mucor*.)

KING, T. J., *Green plants and their allies*. Thomas Nelson, 1983.
(Plant life cycles.)

KROMMENHOEK, W., SEBUS, J., and VAN ESCH, G. J., *Biological structures*. John Murray, 1979.
(Mammalian gonads; angiosperm development; *Capsella* development.)

LEWIS, D., Studies in Biology No. 110, *Sexual incompatibility in plants*. Edward Arnold, 1979.
(Further details of mechanisms to promote cross-fertilization in plants.)

LOWSON, J. M., *Textbook of botany*, 15th edn, revised by SIMON, E. W., DORMER, K. J., and HARTSHORNE, J. N. University Tutorial Press, 1981.
(Plant life cycles; pollination; embryo development, etc.)

MEEUSE, B. and MORRIS, S. *The sex life of flowers*. Faber & Faber, 1984.
(Beautifully illustrated account of pollination mechanisms.)

PROCTOR, M. and YEO, P., *The pollination of flowers*. William Collins, 1973.
(Pollination mechanisms and pollination vectors.)

ROBERTS, M. B. V., *Biology: a functional approach*, 4th edn. Thomas Nelson, 1986.
(Life cycles; seed formation; mammalian gonads; angiosperm development.)

SHARP, J. A., Studies in Biology No. 82, *An introduction to animal tissue culture*. Edward Arnold, 1977.
(Details of how cultures are set up and examined.)
SHAW, A. C., LAZELL, S. K., and FOSTER, G. N., *Photomicrographs of the flowering plant*. Longman, 1965.
(Development of *Capsella*.)
SHAW, A. C., LAZELL, S. K. and FOSTER, G. N., *Photomicrographs of the non-flowering plant*. Longman, 1968.
(Reproduction in the fungi; life cycle of ferns.)
SHEPARD, J. F., The regeneration of potato plants from leaf cell protoplasts. *Scientific American*, **246**(5), 1982, 112–121.
(Tissue culture using cell protoplasts.)
SIMPKINS, J. and WILLIAMS, J. I., *Advanced biology*. Bell & Hyman, 1981.
(Angiosperm embryo development; mammalian gonads.)
SIMPKINS, J. and WILLIAMS, J. I., *Biology of the cell, mammal and flowering plant*. Bell & Hyman, 1981.
(Angiosperm embryo development; mammalian gonads.)
TOOLE, A. G. and TOOLE, S. M., *A-level biology course companion*. Charles Letts Books, 1982.
(Angiosperm life cycle; mammalian gonads.)
WILLIAMS, J. I. and SHAW, M., *Micro-organisms*, 2nd edn. Bell & Hyman, 1982.
(Reproduction in fungi.)